A New True Book

GLOBAL CHANGE

By Theodore P. Snow

Consultant: James J. MacKenzie, Ph.D.
Senior Associate, World Resources Institute

 CHILDRENS PRESS®
CHICAGO

Workers clean up an oil spill
on a Pacific Ocean beach.

PHOTO CREDITS

© Cameramann International, Ltd.—20, 27
(right)

© Victor Englebert—Cover (both pictures),
9 (right), 22, 35 (bottom right), 37 (right)

© Jerry Hennen—9 (left)

Odyssey Productions—© Robert Frerck, 4,
35 (top right)

Photri—31 (left), 33 (left), 41; Briggs, 24;
© Nancy Ferg, 29 (left); © Michael Habichi,
29 (right)

© Lynn M. Stone—7

TSW/CLICK-Chicago—45 (top); © Tom
Dietrich, 21; © Sue Cunningham, 37 (left);
© Syd Brak, 39; © Willard Clay, 45 (bottom
left); © Charles Seabourne, 45 (bottom right)

Valan—© Thomas Kitchin, 2, 30 (right), 31
(right); © Val Wilkinson, 15 (left), 30 (left);
© K. Ghani, 15 (right); © J. A. Wilkinson, 19;
© Steve Kraseman, 25 (left); © Johnny
Johnson, 25 (right); © Francis Lepine, 27 (left);
© Phillip Norton, 33 (right), 35 (left), 43
(2 photos)

Art—© Chuck Hills, 17; © Al Magnus, 11, 12
(2 illustrations)

Cover—Left: Rain forest in Colombia
Right: Rain forest destroyed for
production of charcoal

Library of Congress Cataloging-in-Publication Data

Snow, Theodore P. (Theodore Peck)
  Global change / by Theodore P. Snow.
    p.   cm. — (A New true book)
  Includes index.
  Summary: Describes some of the bad changes man
is inflicting on the world, such as the cutting down of
tropical forests, the destroying of ozone, oil spills, and
acid rain.
  IBSN 0-516-01105-7
  1. Man—Influence on nature—Juvenile literature.
2. Pollution—Juvenile literature. [1. Man—
Influence on nature. 2. Pollution.] I. Title.
GF75.S66 1990   90-37680
363.73—dc20   CIP
                                    AC

TABLE OF CONTENTS

CHANGES LONG AGO

Many people think that the earth never changes. Its oceans, mountains, and continents have been where they are for as long as people have been on earth. But scientists know that the earth does change. It has changed in the past and it will change in the future.

Opposite page: Rock formations on the Hawaiian island of Maui

By studying rocks and
rivers, mountains and seas,
scientists can tell that the
earth was very different
when it was young. The
earth began about 4.5 billion
years ago as a round ball of
rock that was so hot that it
was partly melted. At first
there were no oceans.
Gradually the earth cooled
and the outer layers became
hard and rocky. But it stayed
hot inside.

Oceans cover about 70 percent of the earth.

As the earth cooled, water
was released from the rocks
to form the oceans. Gases
escaped from the inside of
the earth. These gases
formed the atmosphere. At
first, the gases in the air
were different from the

gases in the air today. Then,
about 3.5 billion years ago,
the first forms of life appeared
in the oceans. They were
too small to see, but they
gradually developed into
larger and more
complicated life forms—
plants and animals.

The plants used sunlight
and a gas called carbon
dioxide to make another gas,
called oxygen. Some of the
oxygen formed a gas called

Plants use sunlight and
carbon dioxide to make food,
giving off oxygen in the process.

ozone. A layer of ozone gas
developed high above the
earth. This ozone layer
protected the early life forms
from harmful rays that come
from the sun. Plants and
animals could now live on
land.

Oxygen is the gas that humans breathe. Today, plants still make oxygen from carbon dioxide and sunlight, and we still have enough oxygen to breathe.

The earth has been changing along with the plants and animals. The earth's hard outer layer, called the crust, is made of several huge pieces, or plates, that move around. The dry-land parts of the

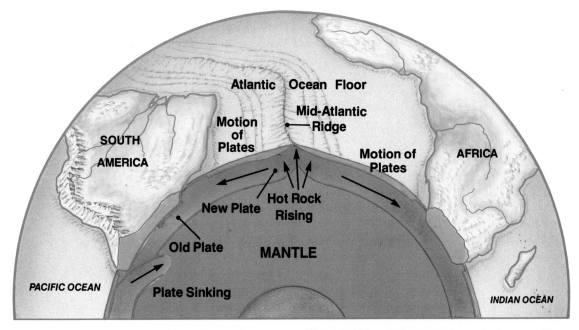

The earth's plates move and change. New material is added to the plates at mid-ocean ridges.

earth, called continents, are
slowly moving along with the
plates they rest on. When the
plates crash together, their
edges crumple up, causing
mountain ranges to form.
There are earthquakes and

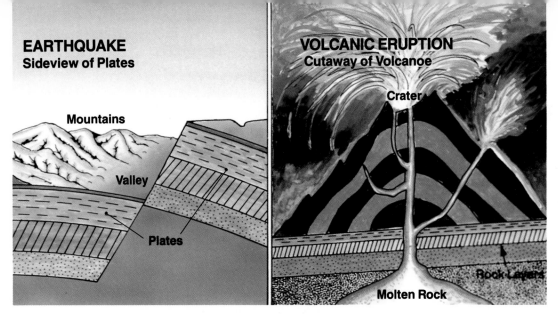

EARTHQUAKE
Sideview of Plates

Mountains

Valley

Plates

VOLCANIC ERUPTION
Cutaway of Volcanoe

Crater

Rock Layers

Molten Rock

Earthquakes (left) happen when the rocky edges of the plates break and shift due to pressure when they slide past each other. Volcanoes (right) erupt when gases and hot, melted rock from inside the earth pour from the volcano's crater.

volcanic eruptions where the plates meet.

There have been great changes in the air and in the earth's landforms since the earth began. Our weather has changed many times as well, for reasons that scientists do not fully understand.

CHANGES CAUSED BY PEOPLE

In the past, all the changes in the earth and its weather were caused by natural forces within the earth and by the natural things that plants and animals do. But today, new kinds of changes are being caused by things that people do. Some of these changes are bad for the earth, making it harder for

people, plants, and animals
to live. Scientists call what is
happening to the earth
global change.

Scientists still do not know
all the things that cause
global change. But some
things are known. We know
that the weather can be
changed by letting certain
gases escape into the air.
We know that the water in
rivers, lakes, and oceans can
be changed by chemicals
that are poured into the
water. We know that some

Trash and chemicals pollute a river in Canada (left). Fish and other water animals are killed when oil is spilled from ships (right).

kinds of plants and animals
have already died out
because people have ruined
the places where the plants
and animals lived. By
building houses to live in
and using machines to make
our lives more comfortable,
we are making life harder for
many other living things.

15

CHANGES IN THE WEATHER

The air that we breathe is made up of gases. Two of these gases are oxygen and nitrogen. There are many other gases mixed in, but only in very small amounts. When sunlight strikes the earth, its energy is changed into heat energy. The earth and the air are warmed by the sun's rays. Any extra heat passes out into space.

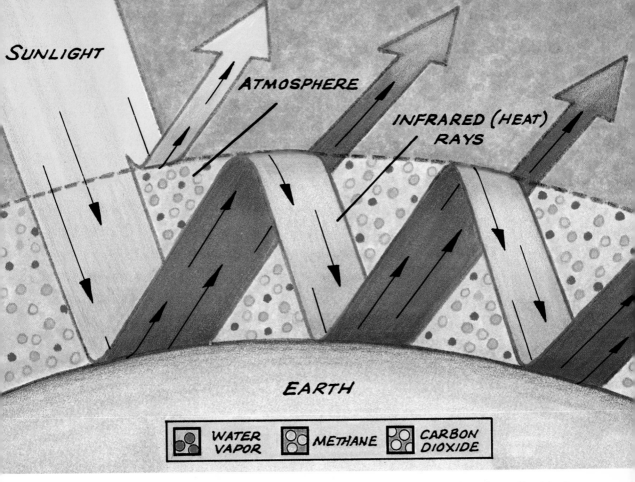

SUNLIGHT

ATMOSPHERE

INFRARED (HEAT) RAYS

EARTH

WATER VAPOR   METHANE   CARBON DIOXIDE

Greenhouse gases such as methane and carbon dioxide keep
the earth's extra heat from escaping into space.

But some of the gases in the
air trap the sun's heat and
make the air warmer. This
heating is called the
greenhouse effect, because
the gases trap heat like the

17

glass in a greenhouse.

One of these gases is methane. This gas is produced by the decay of plants underwater and by cattle and termites when they digest their food.

Another gas that traps heat is carbon dioxide. Today, a lot of extra carbon dioxide is being released into the air when people burn things. Gas, oil, and coal are called fossil fuels because they formed from plants and animals that lived a long time ago and then were buried in the earth. Many factories and almost all cars burn fossil fuels. Fossil fuels release carbon dioxide when they are

Chemicals in the gases from factory smokestacks put greenhouse gases into the air.

The burning of oil products in cars and trucks accounts for about
30 percent of the carbon dioxide we are putting into the air.

burned, so lots of the extra
carbon dioxide in the air comes
from factories and cars.

Burning wood also
releases carbon dioxide. In
South America, people are
burning huge rain forests to

The air in many U.S. cities is polluted by smog—chemicals produced by the action of sunlight on the gases emitted by the burning of fossil fuels.

Bare soil left by the cutting and burning of forests can be washed away by rains.

make more grazing land for cattle to feed on. The fires give off carbon dioxide and other greenhouse gases, such as nitrous oxide.

Burning these forests changes the environment. It

destroys the homes of many animals, and it leaves the soil bare. Without the plant roots to hold it, the soil washes away when it rains. Burning these forests is very bad for the earth.

Scientists are sure that the extra carbon dioxide in the air will make the earth's weather warmer. But they are not sure how much warmer the weather will become. In a few years,

Global warming could cause rainfall patterns
to change. A drought would change productive
farmland into dried-up wasteland.

parts of the world that are
comfortable now might be
too hot, and parts that are
cold now might be warm. It
might become impossible to
grow crops in some areas
where they grow today. Also,
the extra heat may cause

Warmer temperatures could cause the ice in glaciers and ice sheets to melt, raising the level of the oceans.

some of the ice near the North and South poles to melt. This could make the oceans rise higher on the shores. By the year 2100, the oceans could rise three feet, enough to flood some coastal cities.

THE LOSS OF OZONE

Another gas that is present in the air is called ozone. Most of the ozone is formed very high up in the atmosphere—higher than the level where most airplanes fly. The layer of ozone protects us from harmful rays from the sun called ultraviolet light.

Today, some chemicals that are being released into the air are destroying that ozone. These chemicals are

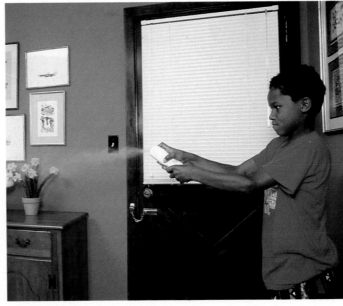

The world's industries (left) used more than 2 billion pounds of CFCs in 1987. Some spray cans contain CFCs (right).

chlorofluorocarbons, or CFCs. They contain a substance called chlorine, which destroys the ozone. CFCs are used in air conditioners, refrigerators, and some spray cans.

27

When CFCs are used, chlorine can be released into the air. Chlorine in the upper air can destroy the ozone layer, allowing more ultraviolet light to reach the earth.

Scientists have found that there is less ozone over the North and South poles of the earth than there used to be. This reduced ozone lets more ultraviolet light reach the ground. The ultraviolet light could cause an increase in skin cancer or eye diseases.

Fish cannot live in polluted
water (left). People (right)
cannot enjoy fishing or swimming
in this polluted stream.

CHANGES IN THE WATER

People, plants, and
animals must have clean
water to drink, and fish must
have clean water to live in.
But now it is getting hard to
find enough clean water in
some parts of the world. The

Some pollution shows up as foam in the water (left).
Barrels of chemicals have broken open (right),
spilling pollutants into the water.

water is polluted by
chemical waste from
factories and by garbage
that people throw away.
Floating trash is found on
the oceans all over the
world, and many rivers and
lakes are so full of chemicals
that fish cannot live in them.

One of the worst things that can be spilled into water is oil. Huge ships carry oil across the oceans, and when one of these ships spills oil into the water, it does great damage to the environment. The water stays dirty for a long time,

The dark streaks on the ocean (left) are caused by oil spilled from a tanker. Birds and other animals die when the oil covers them (right).

and the shoreline may be coated with oil for years. Many sea animals and plants are killed whenever there is a big oil spill.

Heat can make water bad for fish and plants, too. Some factories, especially power plants where electricity is made, pour hot water into streams and rivers. That extra heat can kill fish and water plants.

Even the rain is not always clean. Some gases from factory smokestacks can

Some scientists think that acid rain is destroying forests far from the source of the pollution (left). This spruce tree (right) shows dead needles caused by acid rain.

turn rainwater into a mild acid that can hurt living plants and animals. Acid rain has killed fish and plants in many rivers and lakes. Sometimes the acid rain falls hundreds of miles away from the factories that cause it.

DESTRUCTION OF HABITATS

The place where a plant or animal lives is called its habitat. A habitat may be a forest or a river, a field or an ocean. When a habitat is destroyed, the plants and animals that live in it may die. People have destroyed many habitats by building houses, cities, and factories. Many kinds of plants and animals have become extinct when people have destroyed their habitats.

People in Canada (left) protesting acid rain pollution caused by factories like the one at top right. A valuable rain forest was cut down to make room for this city (bottom right) in South America.

Today, more habitats are being destroyed. Cities all over the world are growing bigger and spreading into the countryside around them. Pollution in the air and in the water is destroying the habitats of many kinds of fish and birds. Oil spills in the oceans have destroyed the habitats of fish and animals that live along the shoreline.

The worst habitat destruction of all is the cutting and burning of rain forests in the tropical parts of the world, such as South America and Africa. No one knows how many different kinds of plants and animals live in these forests. Some of these plants and animals

People are cutting down the rain forests for cattle
ranches (left) and for oil exploration (right).

might give us medicines to
cure diseases, but many of
them may die before scientists
ever find out about them.

Every day, trees in these
forests are cut down and
burned. This is bad in many

ways. Habitats are destroyed, carbon dioxide is added to the atmosphere, and the soil is washed away by rainfall after the trees are gone.

Today, many people want the cutting of the rain forests to be stopped. But the farmers who cut and burn the trees want land for their cattle.

This picture uses a model of the earth to show how CFCs from spray cans destroy the ozone layer.

WHAT WILL HAPPEN NEXT?

A lot of bad things are happening to our earth today. But some good things are starting to happen, too. Maybe the earth will not change so much.

Scientists are not sure how much the things people are doing will affect the earth. Maybe the earth will be able to stay almost the same in spite of pollution and habitat destruction. Maybe some of the changes people are causing will help make the earth better to live in, rather than worse.

But we cannot afford to just wait and see. We must

A scientist keeps records of the amount of ozone high in the air from information sent back to earth by an airplane carrying measurement instruments.

do as much as we can to stop the bad changes from getting worse.

People are at last beginning to learn about global change. This will help us to stop doing things that are bad for the earth. Many factories have stopped letting harmful chemicals get into the air and the water. Many people are recycling their trash instead of just throwing it away.

Governments are trying to stop the destruction of the huge rain forests and the killing of plants and animals. Some rivers and lakes that were once too dirty for fish

Scientists experiment with the effects of acid rain on tree growth (left). The soil in a forest is examined for the effects of acid rain (right).

to live in have been cleaned up and now have as many fish as ever. Many countries have agreed to reduce the production of CFCs.

Perhaps the greatest danger facing earth is overpopulation. More and more people are using energy resources and taking more and more land for farming. Both of these activities add to the problem of global change.

It will take a long time for people all over the world to stop doing things that are harmful for the earth. But maybe, if we all try, we can keep the earth as it is—a wonderful place to live in.

We must all work
together to keep our
beautiful "blue marble"—
the earth—a place
where all living
things can grow
and flourish.

WORDS YOU SHOULD KNOW

acid (ASIHD) — a sour-tasting chemical that can be harmful to
living things

acid rain (ASIHD RAYN) — rainwater that has a high acid content

atmosphere (AT • muss • feer) — the gases surrounding the earth
and some other planets; the air

carbon dioxide (KAR • bun dye • OX • ide) — a gas in the air that
is made up of carbon and oxygen

CFCs (chlorofluorocarbons) (KLOR • oh • FLOO • roh • kar •
bunz) — manufactured gases
containing chlorine and fluorine
that block the sun's rays

chemicals (KEM • ih • kilz) — materials that are used in many
manufacturing processes and that are often harmful to
living things

chlorine (KLOR • een) — a chemical found in CFCs that is harmful
to the ozone layer

continent (KAHN • tih • nent) — a large landmass on the earth

crust (KRUHST) — the top layer of the earth

earthquake (ERTH • kwayk) — shaking of the ground caused
when the earth's plates suddenly move

environment (en • VY • ron • mint) — the things that surround a
plant or an animal; the lands and waters of the earth

extinct (ex • TINKT) — no longer living

fossil (FAW • sill) — the hardened remains of a plant or an animal
that lived long ago

global (GLOH • bil) — covering the whole earth

greenhouse (GREEN • house) — a building with glass or plastic
walls and roof, used for growing plants

habitat (HAB • ih • tat) — the place where a plant or an animal
usually is found

methane (MEH • thayn) — a gas that is produced when plant
matter decays in a place where there is little or no oxygen

nitrogen (NY • tro • jin) —a gas in the air

nitrous oxide (NY • truss OX • ide) —a gas that is produced by burning

overpopulation (oh • ver • pop • yoo • LAY • shun) —too many people for the environment to support

oxygen (OX • ih • gin) —a gas that is found in the air and that humans and animals need to breathe

ozone (OH • zohn) —a gas high in the air that protects the earth from ultraviolet light

rain forest (RAYN FOR • ist) —a thick evergreen forest that grows in the tropics, where there is much rain

recycling (re • SY • kling) —using something again; breaking down a product into its parts for reuse

tropical (TRAH • pih • kil) —found near the equator, an imaginary line around the earth halfway between the North and South poles

ultraviolet light (ul • tra • VYE • lit) —rays from the sun that are invisible and that can cause harm to living things

volcanic eruption (vahl • KAN • ik ee • RUP • shun) —melted rock and gases from deep inside the earth bursting out of a volcano

INDEX

About the Author

As a child, Theodore P. Snow read science books and science fiction stories and became interested in astronomy. He majored in astronomy and physics at Yale University and received graduate degrees from the University of Washington in Seattle. In the 1970s, Dr. Snow became involved in the space program through his work with the space telescope at Princeton University. Since 1977, he has taught astronomy at the University of Colorado at Boulder. Dr. Snow and his wife, Connie, have three sons, who are also space enthusiasts.